THE EARTH

By
Juhi Shrivastava

© 2020 Juhi Shrivastava

ISBN:9789353966546

Contents

THE EARTH

Do You Know!!

❖Planet Earth is third planet from Sun.

❖Earth was formed over 4.6 Billion years ago.

❖Earth is only planet in solar system which is in habitable zone for life to flourish.

❖Around 29% of earth surface consists of land mass and remaining 71% consists of water.

The Big Blue Marble we lovingly call our home has average radius of
 6378.1 kilometers.
It is home to more than 7.5 billion humans.

THE EARTH'S PLACE IN SOLAR SYSTEM

❖The Earth along with other planets of Solar System revolve around Sun.

❖The path of Earth around sun is called its Orbit.

❖One orbit of Earth around sun takes 365.256 days.

❖Earths orbits at a distance of 149.6 million kilometers or 92.96 million miles from the Sun.

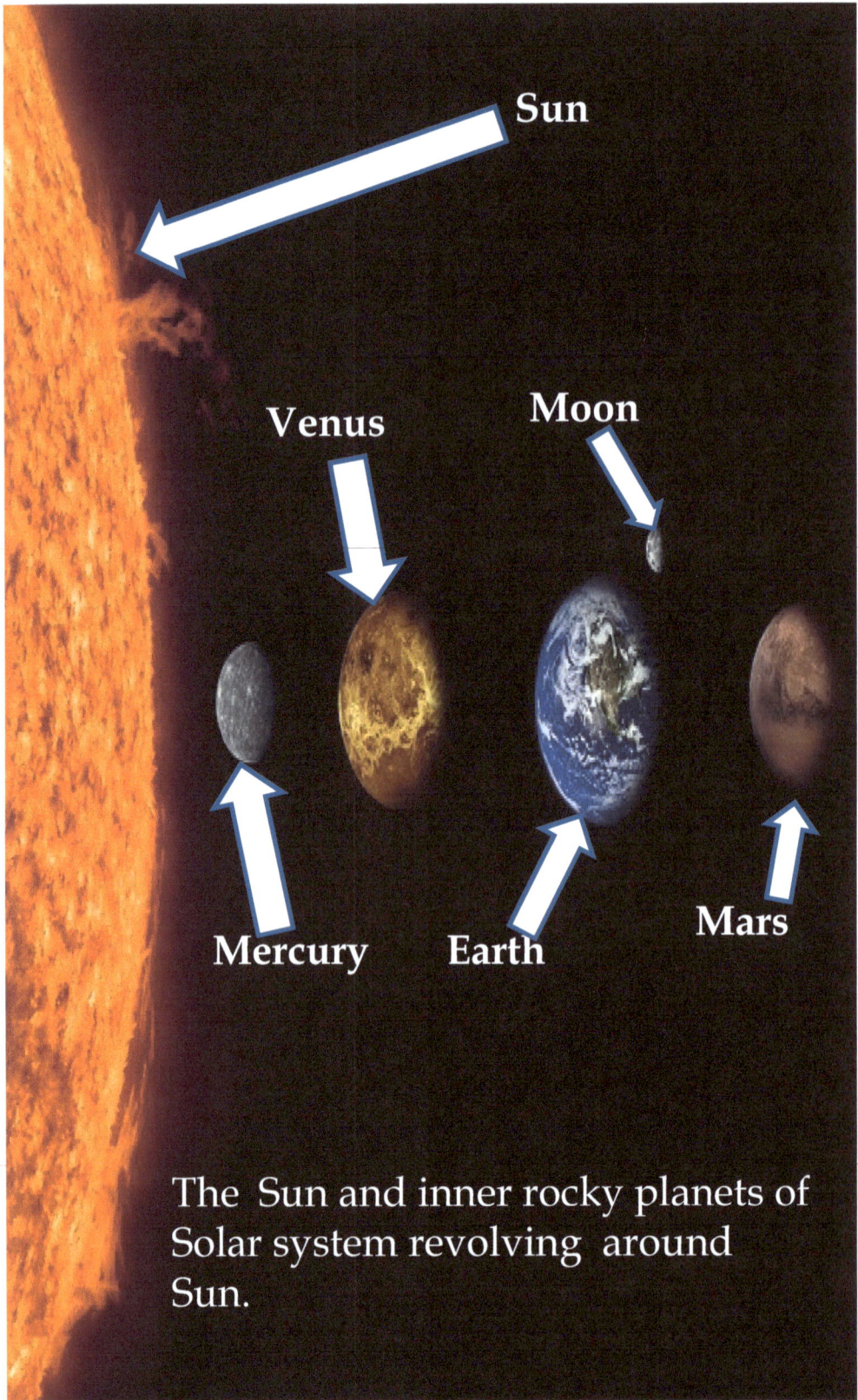

Sun

Venus

Moon

Mercury

Earth

Mars

The Sun and inner rocky planets of Solar system revolving around Sun.

THE EARTH FROM SPACE

❖Although all the human beings live on Earth, very few of us have got chance to see Earth from outside.

❖To see Earth from outside you need spacecraft to fly out of Earth.

❖The most famous picture taken of Earth from outside is called Earthrise.

❖Earthrise is the picture of earth taken from moon on 24th December 1968 by Apollo 8 Astronaut William Anders.

Picture of Earth taken from Moon on 24th December -1968 by Apollo 8 Astronaut William Anders.

THE EARTH'S MOON

❖Earth has one Moon.

❖The Moon of Earth is 384,400 kilometres or 238,885 miles away from Earth.

❖The Moon circles around the earth and its path is called orbit.

❖The Moon is a satellite of Earth ,it is not a planet.

The Moon Takes 28 days to travel around Earth.
The Moon has no air and no life

THE SIZE OF EARTH'S MOON

❖Average diameter of Earth is 12,742 Km and average diameter of Moon is 3,474 Km.

❖Surface area of Earth is 510 million Square Km and surface area of Moon is 37 .9 million square Km

❖Mass of Earth is 5.97 X 10^{24} Kg and Mass of Moon is 7.347 X 10^{22} Kg.

❖Mass of Moon is only 1.2% of Earth's mass.

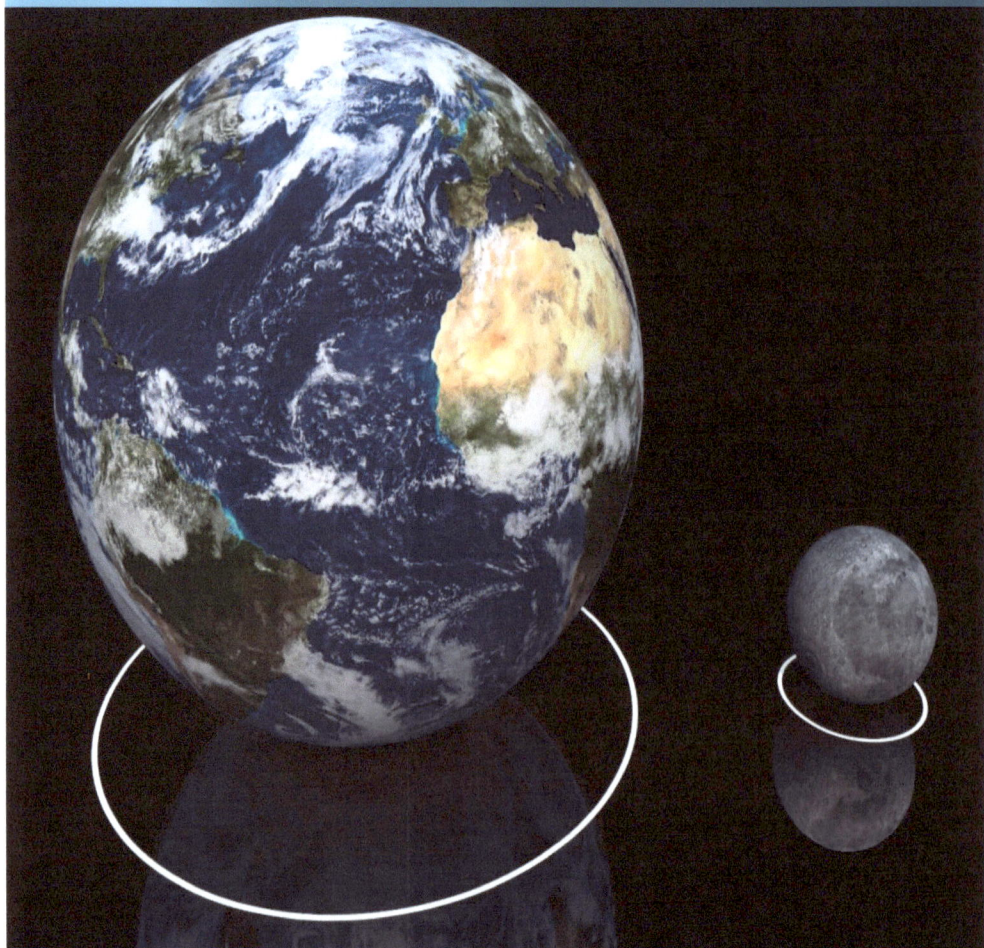

THE EARTH AND THE EARTH'S MOON

THE EARTH FEATURES

❖The Earth is alive planet with Plate tectonic changing the planet surface continuously creating new mountains and new oceans.

❖Himalayan mountain belts are the youngest and tallest mountain belts on Earth. Himalayan mountain belt was formed around 50 million years ago.

❖Atlantic ocean is the youngest ocean formed around 180 Million years ago.

Mount
Everest

Mount Everest the tallest point on Earth is in Himalayan ranges. Mount Everest stands at 8,848 meters above mean sea level

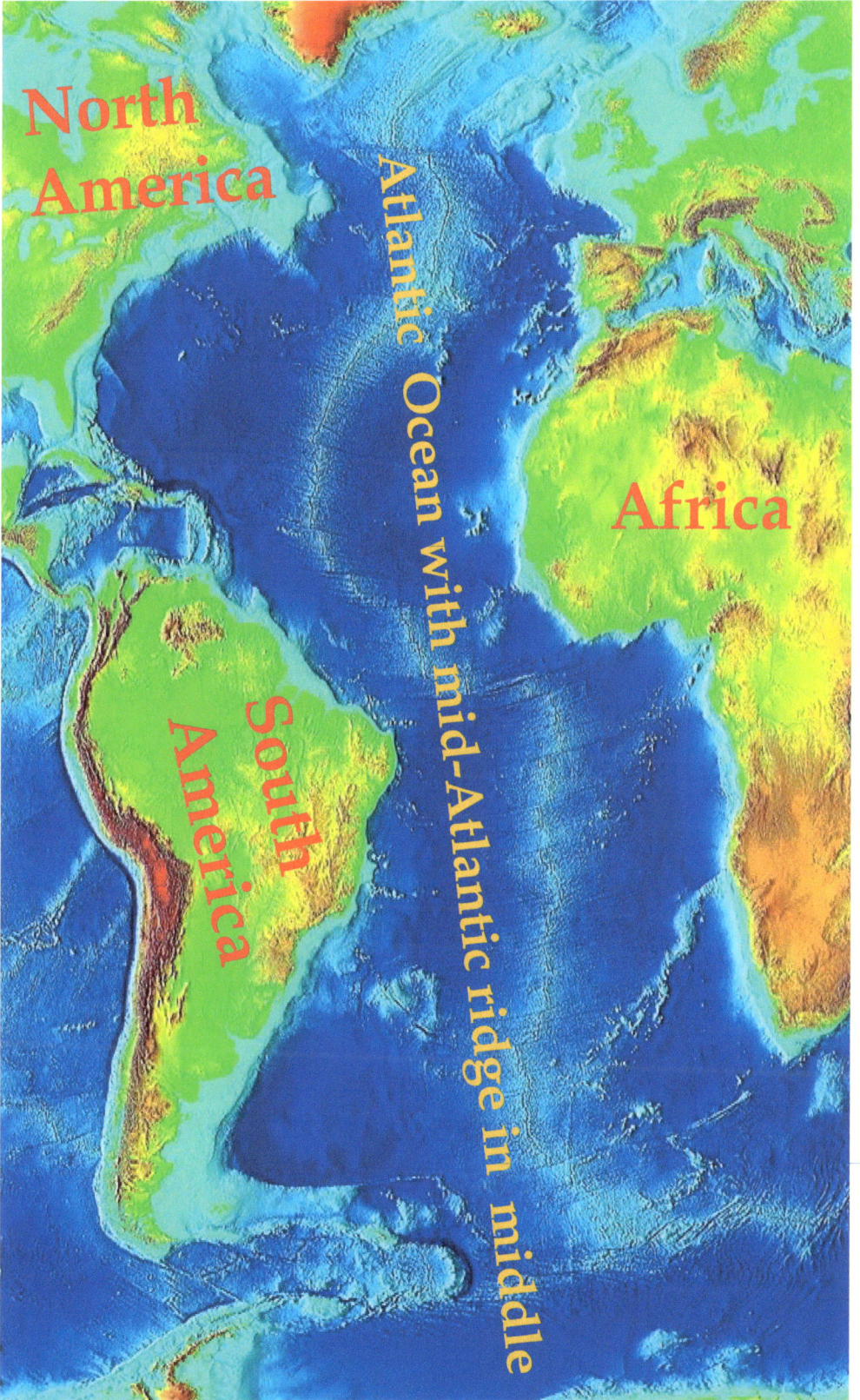

North America

Africa

South America

Atlantic Ocean with mid-Atlantic ridge in middle

OCEAN'S ON EARTH

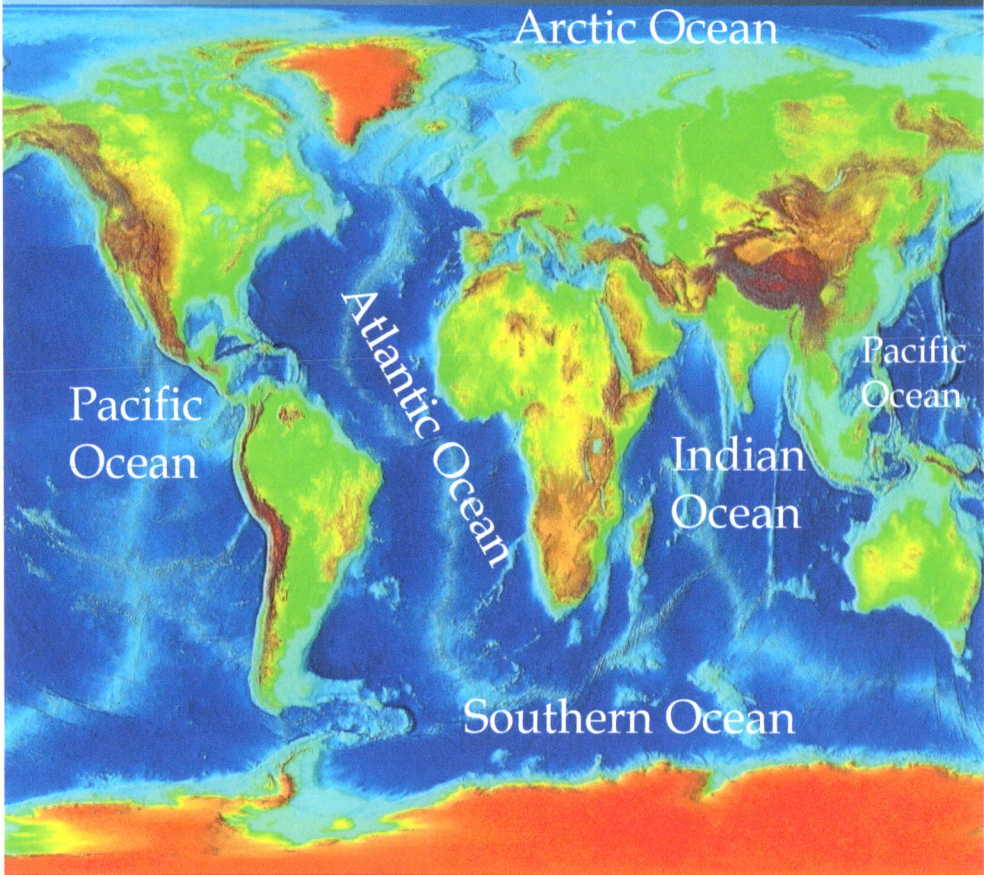

Arctic Ocean

Atlantic Ocean

Pacific Ocean

Pacific Ocean

Indian Ocean

Southern Ocean

There are five Oceans on Earth:
1. Pacifc Ocean
2. Atlantic Ocean
3. Indian Ocean
4. Southern Ocean
5. Arctic Ocean

PACIFIC OCEAN

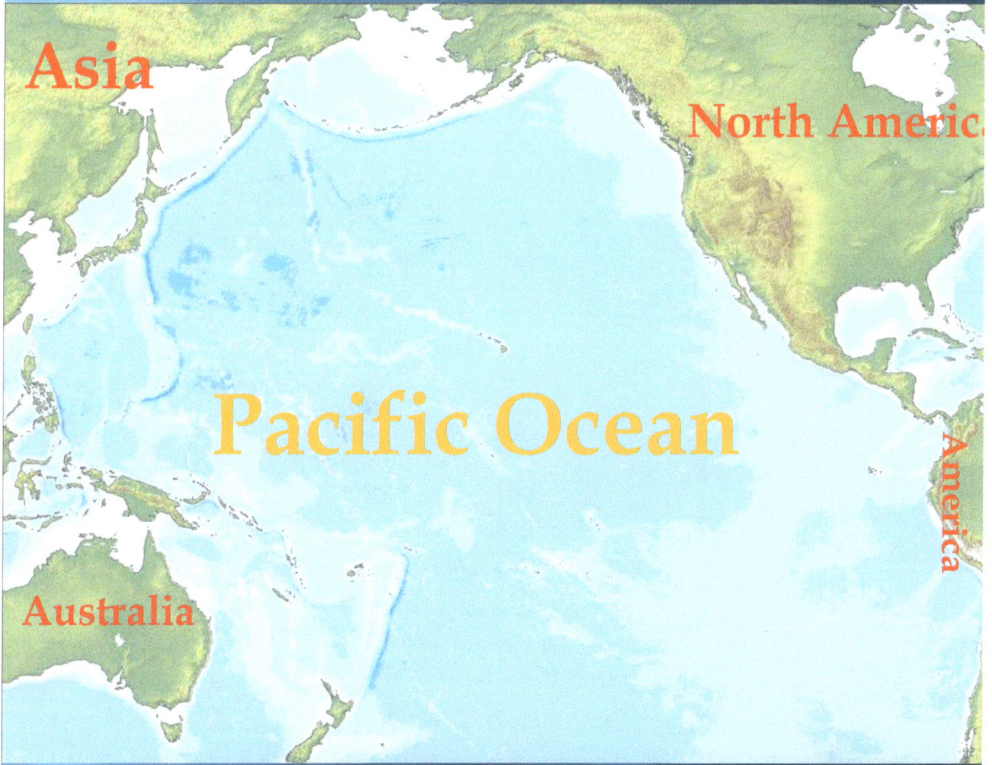

Asia

North America

Pacific Ocean

America

Australia

Pacific ocean is the largest ocean present on Earth.

ARCTIC OCEAN

❖Arctic ocean is the smallest , shallowest and coldest ocean present on Earth.

❖Arctic ocean lies on Earth's north pole.

RING OF FIRE

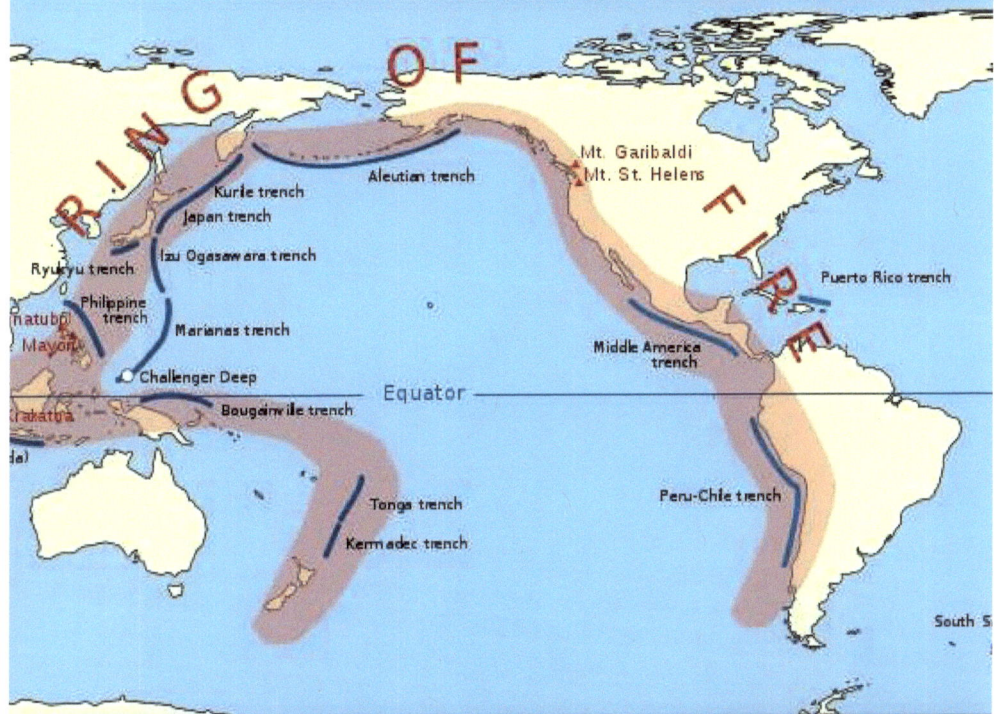

❖Ring of Fire is a major part of Pacific ocean where numerous volcanoes and earthquakes occurs.

❖It comprises 452 volcanoes which is more than 75% of worlds volcanoes.

❖Also around 90% of all the Earthquakes happen in Ring of Fire.

MARIANA TRENCH

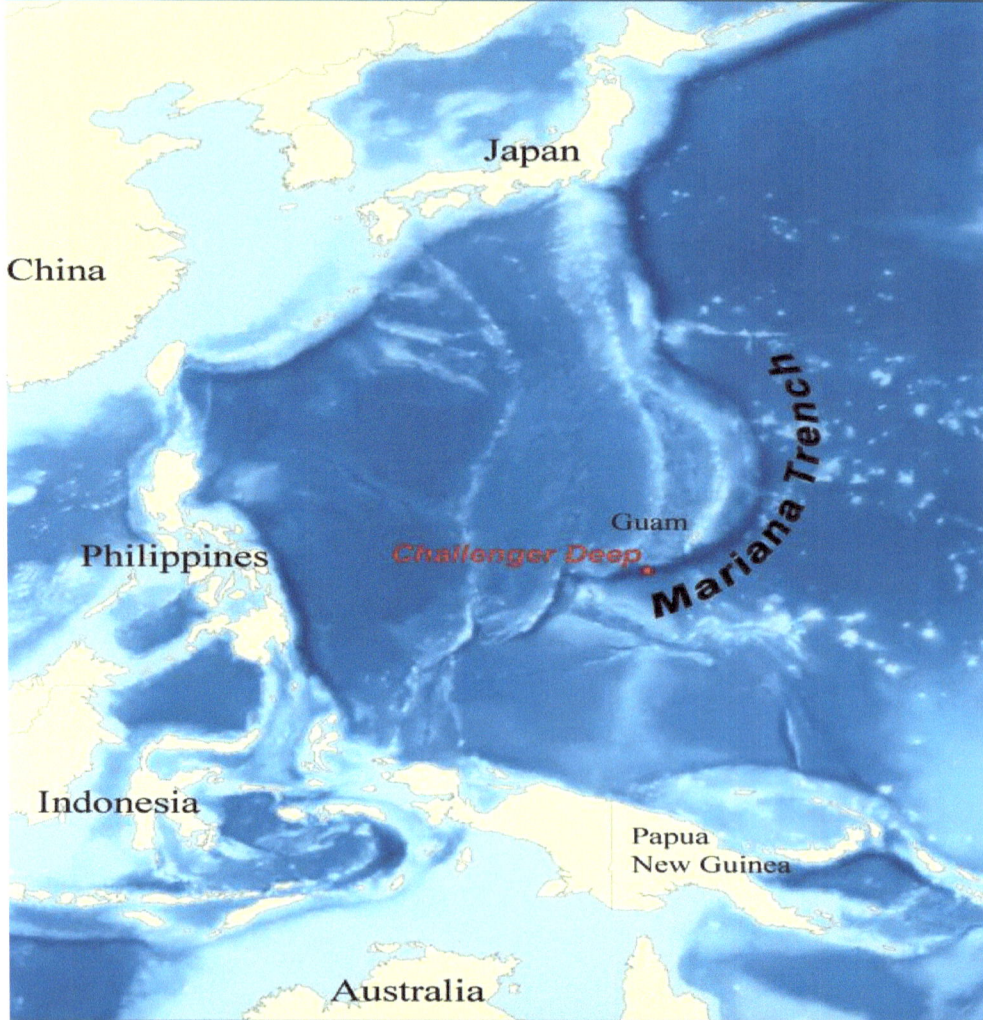

❖Deepest point on Earth "Challenger Deep" is a part of Mariana Trench.

❖Mariana Trench is located in Pacific ocean.

❖Depth of Challenger Deep is 10,984 metres (±25metres)

CONTINENTS

Large Landmasses present on Earth are called Continents.

There are seven continents on Earth . List of continents ordered from largest in area to smallest in area-
1. Asia
2. Africa
3. North America
4. South America
5. Antartica
6. Europe
7. Australia

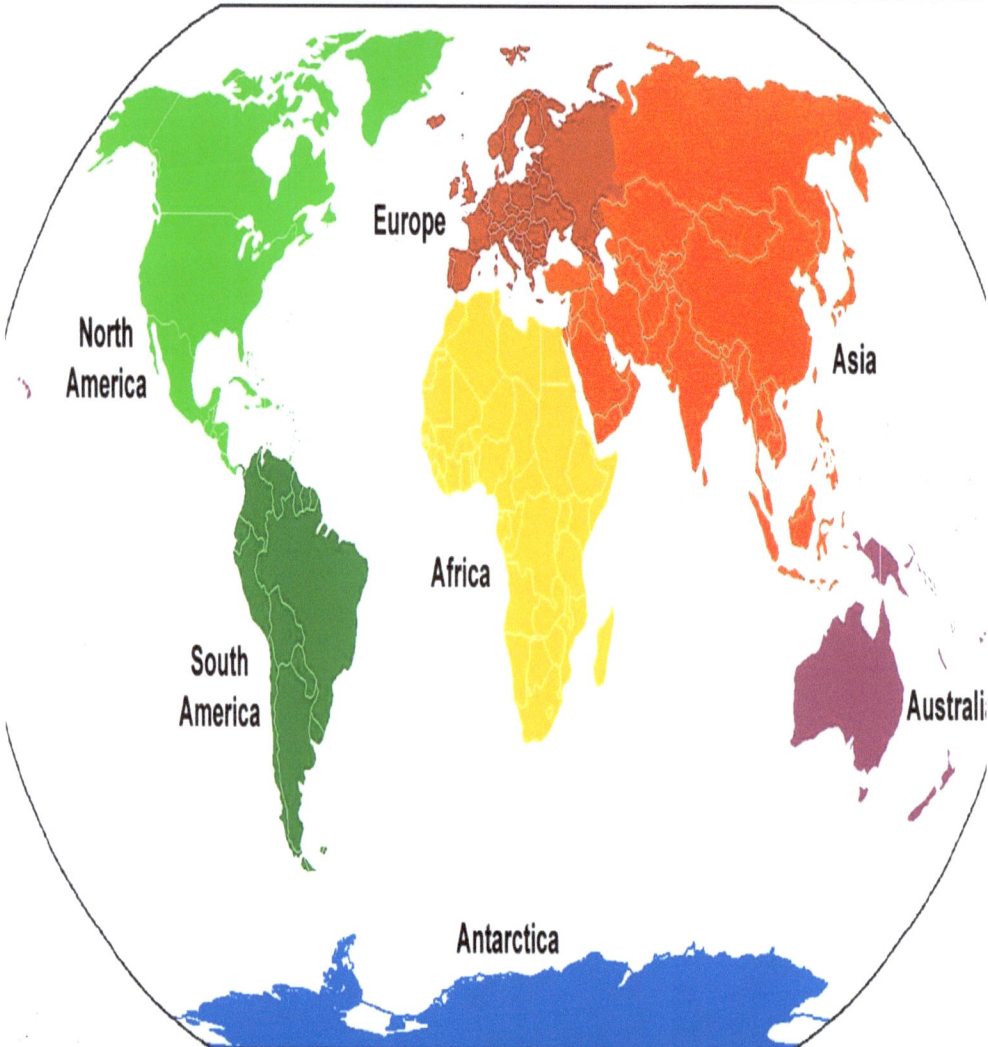

NEW WORDS YOU LEARN

❖Planet : Round astronomical body that has cleared its neighbouring region of plantesimals.

❖Orbit: The path of an object around another object.

❖Satellite: An astronomical body that orbits a larger planet.

❖Mountain :Large mass of land which rises above surrounding landmass.

❖Ocean: Large body of water which has most of the planets water.

❖Continent: Large mass of land on planet.

The Earth by Juhi Shrivastava

**Children's book about the Earth with Fun facts & pictures.
For kids of age 5 to 8 years.**

www.ingramcontent.com/pod-product-compliance
Lightning Source LLC
Chambersburg PA
CBHW041719200326
41520CB00001B/168